EXPOSURE REGISTERS IN EUROPE

Extractions of core information
and possibilities
for comparison between
European databases for occupational
air pollutant measurements

European Foundation
for the Improvement of
Living and Working Conditions

EF/94/22/EN

EXPOSURE REGISTERS IN EUROPE

Extractions of core information
and possibilities
for comparison between
European databases for occupational
air pollutant measurements

by
**P. Vinzents, B. Carton, P. Fjeldstad,
B. Rajan, R. Stamm**

European Foundation
for the Improvement of
Living and Working Conditions
Loughlinstown House, Shankill, Co. Dublin, Ireland.
Tel: +353 1 282 6888   Fax: +353 1 282 6456

Cataloguing data can be found at the end of this publication

OBN 1216617

Luxembourg: Office for Official Publications of the European Communities, 1994

ISBN 92-826-8737-6

© European Foundation for the Improvement of Living and Working Conditions, 1994

For rights of translation or reproduction, applications should be made to the Director, European Foundation for the Improvement of Living and Working Conditions, Loughlinstown House, Shankill, Co. Dublin, Ireland.

Printed in Ireland

PREFACE

Collecting and analysing data on health and safety at the workplace is an essential starting point for setting up priorities and action plans as well as for evaluation of the measures taken.

Since 1988 the Foundation has been working on this topic at a European level in order to seek a more co-ordinated approach to the monitoring of working conditions relating to health and safety. A number of European networks and working groups have been established in order to assist the Foundation in this process.

One of the networks which exchange information on "hard data" on health and safety concerns "Exposure Register in Europe".

In this publication the network is reporting on the possibilities for comparison between the databases exemplified by xylene measurements.

Pascal Paoli　　　　　　　　　　　　　　　　　　Henrik Litske
Research Manager　　　　　　　　　　　　　　*Research Manager*

FOREWORD

The Foundation established the European Network of Exposure Registers in 1991.

This report is proposed by a working group of:

P. Vinzents	National Institute of Occupational Health, Copenhagen, Denmark
B. Carton	Institut National de Recherche et de Sécurité, Vandoeuvre, France
P. Fjeldstad	National Institute of Occupational Health, Oslo, Norway
B. Rajan	Health and Safety Executive, Bootle, UK
R. Stamm	Berufsgenossenschaftliches Institut für Arbeitssicherheit, St. Augustin, Germany

TABLE OF CONTENTS

	Page
Preface	5
Foreword	7
Summary	11
Introduction	13
Presentation of the databases	15
Differences between exposure assessments	26
Presentation of xylene results	30
Representativeness of xylene results	35
Definition of core information	37
Recommendations	38
References	39

SUMMARY

Exposure Registers in Europe. Extraction of core information andpossibilities for comparison between European databases for occupational air pollutant measurements.

by

P. Vinzents, B. Carton, P. Fjeldstad, B. Rajan, R. Stamm.

A study has been performed on five European databases from Germany, France, Norway, United Kingdom and Denmark. The databases contain occupational air pollutant control measurements and are administered by insurance institutions or governmental bodies. The purpose for the study was to discuss the possibilities of making comparisons between the bases and to define the core information which is essential for comparisons.

The discussion was actually carried out on a comparison of xylene measurements stored in the bases. Four classes of xylene measurements are compared:
1: all xylene measurements,
2: xylene measurements within the wood working industry,
3: xylene measurements during spray painting,
4: xylene measurements during spray painting in the wood working industry.

When the mean values are compared no pattern emerges; the spray painting results (3) are elevated compared to the other classes, and (2) and (4) are at the same level as (1). For two of the bases, the standard deviation is decreasing from (1) to (4).
The comparison was carried out under two assumptions: exposure time equal to sampling time, and exposure time equal to 8 hours.
The conclusion come to was that comparisons can now be carried out at industry level, because codes classifying different industries can be translated or compared between countries.

Exposure time and code systems, classifying measurements according to industry, work processes and other categories, are defined as core information. Sampling strategy is another important subject that should be considered when measurements are compared.

In one of the bases a representative unbiased dataset is stored. This dataset is used to quantify the bias caused by the strategy used for control measurements. The result is that the difference between the representative dataset and the control measurements is small.

Finally it is recommended that the co-operation should be continued, and that a future exercise should be carried out with substances which are different from xylene, such as:
- styrene in glass fibre reinforced plastic production,
- lead in battery production,
- percholoethylene in cleaning of clothes.

Comparisons between these substances would facilitate a more detailed discussion on the code systems, because these substances are used in very specific work operations.

SUMMARY

Exposure Registers in Europe. Extraction of core information andpossibilities for comparison between European databases for occupational air pollutant measurements.

by

P. Vinzents, B. Carton, P. Fjeldstad, B. Rajan, R. Stamm.

A study has been performed on five European databases from Germany, France, Norway, United Kingdom and Denmark. The databases contain occupational air pollutant control measurements and are administered by insurance institutions or governmental bodies. The purpose for the study was to discuss the possibilities of making comparisons between the bases and to define the core information which is essential for comparisons.

The discussion was actually carried out on a comparison of xylene measurements stored in the bases. Four classes of xylene measurements are compared:
1: all xylene measurements,
2: xylene measurements within the wood working industry,
3: xylene measurements during spray painting,
4: xylene measurements during spray painting in the wood working industry.

When the mean values are compared no pattern emerges; the spray painting results (3) are elevated compared to the other classes, and (2) and (4) are at the same level as (1). For two of the bases, the standard deviation is decreasing from (1) to (4).
The comparison was carried out under two assumptions: exposure time equal to sampling time, and exposure time equal to 8 hours.
The conclusion come to was that comparisons can now be carried out at industry level, because codes classifying different industries can be translated or compared between countries.

Exposure time and code systems, classifying measurements according to industry, work processes and other categories, are defined as core information. Sampling strategy is another important subject that should be considered when measurements are compared.

In one of the bases a representative unbiased dataset is stored. This dataset is used to quantify the bias caused by the strategy used for control measurements. The result is that the difference between the representative dataset and the control measurements is small.

Finally it is recommended that the co-operation should be continued, and that a future exercise should be carried out with substances which are different from xylene, such as:
- styrene in glass fibre reinforced plastic production,
- lead in battery production,
- percholoethylene in cleaning of clothes.

Comparisons between these substances would facilitate a more detailed discussion on the code systems, because these substances are used in very specific work operations.

INTRODUCTION

A recent study has identified 41 organizations in Europe holding databases on occupational air pollutants[1]. The 5 databases in this present study alone contain more than one million results of occupational air pollutants. This represents a huge potential of knowledge on occupational safety and health, but it is not fully utilized internationally, because the data are mainly used by national authorities and are not extensively used for scientific purposes.

At the moment, an intensive discussion is going on in the European Community concerning the use of occupational air pollutant measurements, because efforts are being made to establish occupational exposure limits at a European level. In order that decision-makers can exploit existing data on exposure, and their implications for occupational health within the EC, it is important to provide comparisons and analyses of occupational air pollutant measurements.

One precondition for a valid comparison of measurements of occupational air pollutants is that classification systems (codes), sampling strategies, and analyses should be comparable. The aim of this project is therefore to analyse in detail the information needed when comparisons are carried out and, as an example, to compare measured concentrations of a chemical substance.

The group decided to work on xylene (all isomers), because xylene is a common organic solvent, used in many industries. Furthermore a specific industry (the wood and furniture industry) and a specific work operation (spray painting) were chosen. Spray painting was selected because it was assumed that spray painting techniques in the participating countries are similar. Therefore, differences between the bases would be caused by the different measurement strategies on which this project focuses.

Preliminary results from the project have been presented at meetings in General Directorate V, and at two conferences, (i) on occupational exposure databases held by the American Conference of Governmental Industrial Hygienists, November 1993 in Washington DC[2], (ii) at the British Occupational Hygiene Society Conference, March 1994[3].

Research on quantitative occupational exposure assessment deals with categories as homogenous exposure groups and occupational groups[4, 5, 6]. These categories are important in epidemiology and generally in exposure assessment exercises. Unfortunately, however, promising results which would help in establishing these groups do not yet exist. Therefore, one purpose for a new project could be to examine the possibilities for establishing these categories from existing data in databases, and to discuss requirements for future data collection.

PRESENTATION OF THE DATABASES

In this section some information on the databases is given, including:

- administrative systems,
- data sources,
- number of measurements and analyses,
- number of analyzing laboratories,
- codes and other additional information,
- quality control,
- measurement policy,
- sampling strategy,
- exposure assessment.

No technical details on the bases are given, since all the bases are run on modern computer systems where export of any data in a common format is default. At the end of the section, a comparative list of information in the bases is given.

ATABAS, Denmark

The Danish base is administered by the National Institute of Occupational Health (in Danish: AMI), which is part of the Danish Working Environment Service (DWES). The measurements stored in the bases are carried out only by the DWES in all 14 counties in Denmark. All analyses are carried out at AMI. The base contains 70,000 analyses from 18,000 samples.

All measurements are classified using a 4-level hierarchical code system. The levels are:
- industry   (ISIC code),
- department   (ISIC code),
- job   (national code, based on ILO system),
- work operation   (alphabetic list).

Additional information on sampling time and period, the measured persons, protective equipment, exposure time and ventilation are also stored.

The quality of the chemical analyses is supervised by participation in inter-laboratory control programmes (NIOSH, EC BRC and Scandinavian). The additional information in the base is checked by one experienced occupational hygienist.

The measurements are performed by local occupational hygienists and are mainly used by local inspectors to check the air pollutant levels as compared to the occupational exposure limits. The overall sampling strategy thus tends to be biased towards a worst case strategy.

No formal regulations on sampling strategy and sampling time are given, only recommendations based on CEN-work.

Only measurements from personal sampling, where no respiratory protection device was used, are used for exposure assessment. When more than one sample has been taken from one person during the working day, the exposure assessment is carried out as calculation of time weighted average concentration from the dose (mass of pollutant) from each sampling device within the person.

EXPO, Norway

The Norwegian base is administered by the National Institute of Occupational Health in Oslo, a governmental research institute in the fields of occupational medicine, hygiene, physiology and toxicology. It collaborates closely with the Labour Inspection team. The institute is used as a regional laboratory, and receives work environment samples from Norwegian companies for chemical analysis. The samples contain solvent vapour, blood, urine, aerosols etc.

Since 1985 about 65,000 samples (200,000 concentration values) have been analysed at the institute and stored in the database EXPO. At present EXPO covers mostly the eastern part of Norway as far as air samples are concerned, while biological samples to monitor the work environment are received from all over the country. The Labour Inspection team and the Institute of Occupational Health aim at making EXPO cover exposure data from most kinds of work from the whole country.

The measurements are classified according to 4 different codes:
- Industry (ISIC code and an alphabetic list)
- Department (alphabetic list)
- Work operation (text)
- The reason for the measurement (EXPO specific code)

The institute participates in inter-laboratory control programs. The information in EXPO is checked by experienced occupational hygienists. Most of the registration work is done by the institute secretary. Some of the analytical results are transferred electronically from the analytical instruments. The results may be typed in by the analyst himself, who will also decide when an analytical series is finished and should be marked so that the data can no longer be altered.

The employers are responsible for carrying out the necessary sampling and analyses to be sure they are in compliance with legislation and rules for the work environment. The sampling will most often be done by a more or less trained occupational hygienist from a local health service centre. Occasionally, an employer may not use a competent occupational hygienist to obtain the sample. The base registers what kind of institution is responsible for the sampling. At present most of the samples stored in the base are analysed at the institute.

No formal regulations on sampling strategy and sampling time are given.

COLCHIC, France

The COLCHIC database is administrred by the INRS, which works with C-NAM (National Health Insurance Fund) and the local branches, C-RAM. Data are provided by 8 Regional Health Insurance Funds (laboratoires regionaux) which conduct local sampling and analyses. This is also the case for surveys made by the INRS itself. The base contains approximately 205,000 analyses from about 88,000 air samples.

Each measurement is qualified by a set of data, among them:

- industry (social security code)
- workplace
- object
- ventilation qualification
- time and period of sampling
- sampling method.

There is a quality assurance process inside the (8+1) laboratories with exchange of information, and participation in inter-laboratory comparisons (home system, WASP).

The measurements are performed by local engineers, whose main aim is to control the exposures. The picture is probably biased towards a worst case strategy.

MEGA, Germany

The German MEGA base is designed and operated by the Central Institute for Research and Testing of the German Statutory Accident Prevention and Insurance Institutions in Industry (Berufsgenossenschaftliches Institut für Arbeitssicherheit - BIA).

The inspectors of these insurance institutions take samples of the air at a wide variety of workplaces in their member firms.

These samples are forwarded to the BIA, together with characteristic data collected at the workplace. The BIA records all incoming samples and operational data, carries out chemical analyses and draws up an analysis report, which contains all the information needed for an assessment of industrial hygiene at the workplace.

All workplace and measurement data are stored in the MEGA base. Since 1972 MEGA has stored data from 600,000 analyses of organic dusts, gases and vapours, mineral dust and fine dusts from 280,000 samples. Today up to 150 pieces of information are recorded on each measurement e.g. firm, sector, workplace, activity, production process, working materials, physical environment, exposure time, sampling conditions, measuring techniques and measured values.

The information contains individual text and a classified code system. The levels of the code system are:

- branch of industry: national code of the industrial area,
- workplace: working process (BIA code),
- activity: national code of the professional activity.

The MEGA database cannot be regarded as representative in a statistical sense, since firms and inspection sites are selected by the Insurance Institution, not by reference to statistical criteria but as the professional need arises. Firms where increased exposure is expected are therefore likely to be preselected.

To ensure a consistently uniform procedure and to collect exposure data of a uniformly high level of quality; appropriate actions are taken, e.g.:

- the uniform application of validated measuring techniques,
- the provision of uniform working aids,
- plausibility checks and double data input,
- uniform reporting on the findings of analyses, the operational and analytical data being linked to the relevant legislation,
- regular exchanges of experience and suitable training measures,
- participation in multicentre trials to monitor the quality of analyses.

The sampling strategy is determined in the national regulation "Determination and Assessment of Pollutant Concentrations in Workplace Air, TRGS 402".

In accordance with this regulation, the sampling time may be shorter than the exposure time, provided that the sampling time is a part of the exposure time and that the concentration of air pollution in this time is representative for the whole exposure period. The shortest sampling time for an 8 hour exposure period is 2 hours.

NEDB, UK

The UK, National Exposure DataBase (NEDB) is an electronic data system administered by the Technology and Health Sciences Division (THSD) of the Health and Safety Executive (HSE). The HSE is a government body responsible for enforcing the health and safety laws. The NEDB contains exposure information which has been collected by the HSE during the course of special surveys, investigations and inspection of workplaces. The objective of the NEDB is storage, retrieval and analysis of workplace exposure information for substances hazardous to health. One benefit of this database is the ready availability of historical data for workplace health risk management.

The data base is a menu-driven system. A standard proforma is used for the collection of exposure information, which includes the following:

Administrative information - details of the premises, the reason for data collection, visit date, report reference etc.;
Process information - Industry (ISIC code), Process (ISIC code), Job/task (national practice);
Sampling information - whose exposure was measured, agent(s) monitored, for how long, sampling and analytical methods and results;
Control information - comments about the nature of exposure, types of control measures in use, whether Respiratory Protective Equipment (RPE) is used.

Eight laboratories are involved in sampling and analysis. These laboratories participate in national quality control programmes such as Workplace Analytical Scheme for Proficiency (WASP) and Regular Inter-laboratory Counting Exchange (RICE). The data input, retrieval and analysis are managed by qualified Occupational Hygienists. When a new set of exposure information is added to the database, the information is stored in a "temporary file", which is then checked by two other people. Authority to transfer the information to the main file is given by the Occupational Hygienist, who is the database manager. Access to the database is controlled by passwords at all levels.

The database contains over 200,000 results of more than 350 substances. The exposure results are held in two formats - results obtained for the sampling period and 8-h Time Weighted Average (TWA). The calculation of 8-h TWA is carried out following national/international protocols.

Given below is a comparative list of information in the bases. The information is arranged at different levels (x: stored, (x): optional stored, o: not stored):

Database:	DK	N	F	D	UK
enterprise					
name and address of the enterprise	x	x	x	x	x
industrial code	x	x	x	x	x
employer's number	x	x	x	x	x
total number of employees	o	o	o	x	o
description of the workplace					
branch/department	x	x	o	x	x
workplace	x	x	x	x	x
workplace/detailed information	o	x	o	(x)	o
individual workplace/number	o	o	o	(x)	o
number of persons working	o	o	x	(x)	o
number of exposed persons	x	o	x	x	x
respiratory protection	x	x	x	(x)	x
exposure source	o	x	x	(x)	x
profession/jobcode	x	x	o	x	x
identification of person being measured	x	x	x	(x)	x
description of the room					
room/nature	o	o	x	x	x
room/dimensions	o	o	x	x	o
room/open - closed	x	o	x	x	o
free ventilation	x	o	x	x	o
mechanical ventilation	x	o	x	x	o
air conditioning	o	o	x	x	o
suction facility/local exhaust ventilation	x	o	x	x	x
clean air ventilation	o	o	x	x	o
heat recovery	o	o	o	(x)	o
air humidity/pressure	x	o	o	(x)	o
temperature inside	x	o	x	x	o
temperature outside	x	o	o	x	o
weather	x	o	x	(x)	o
wind/direction and intensity	x	o	o	(x)	o
concentration monitoring	o	o	o	(x)	o

24

manufacturing techniques
manufacturing technique/name	o	o	o	(x)	o
manufacturing technique/throughput	o	o	o	(x)	o
manufacturing technique/processing temp.	o	o	o	(x)	o
manufacturing machine/manufacturer	o	o	o	(x)	o
manufacturing machine/year of construction	o	o	o	(x)	o
manufacturing machine/number at the workplace	o	o	o	(x)	o
working manner	o	o	o	(x)	x
working manner/cycles per day	o	o	o	(x)	o

working materials
chemical class	o	o	o	(x)	o
trade name	x	o	o	x	o
manufacturer/name	o	o	o	(x)	o
manufacturer/address	o	o	o	(x)	o
processing capacity	o	o	o	x	o
components	o	o	o	x	o
state of aggregation	o	o	o	(x)	o
skin protection	o	o	o	(x)	o
risk class symbol	o	o	o	(x)	o
safety data number	o	o	o	(x)	o

sampling conditions
reason for measurement	x	x	x	x	x
representative character of the exposure	x	x	x	x	x
sampling date	x	x	x	x	x
sampling time	x	o	x	x	x
sampling duration	x	x	x	x	x
sampling/personal or stationary	x	x	x	x	x
exposure duration	x	x	x	x	x
sampling method	x	x	x	x	x
type of sampling material	x	x	x	x	x
air flow rate (volume)	x	o	x	x	x

analytical conditions
analytical method	x	x	x	x	x
CAS-number/chemical agent	x	x	x	x	x
result	x	x	x	x	x

DIFFERENCES BETWEEN EXPOSURE ASSESSMENTS

As an introduction to the following section, containing the xylene results, we present here some important concepts used in exposure assessment.

Differences in reported occupational exposures to air pollutants may be due to national differences in various categories:
- industrial structure,
- measurement policy,
- sampling strategy.

Differences in industrial structure may lead to differences in measurement policy: the presence of specific industries may lead to development of a national measurement policy aimed at the control of specific substances. These differences reflect real differences in the population's exposure.

On the other hand any differences in sampling strategy may cause different results of exposure assessment, even though the strategies are used on the same population. Sampling strategy may be understood in a statistical way as the process of taking a sub-sample (of air, time, workers, or enterprises) from an overall population. An example of a well-defined sampling strategy is the one prepared by the CEN[7], which compares workers' exposure with the relevant exposure limit values for chemical agents in the air at the workplace. The strategy is designed for exposure assessment at specific workplaces for control purposes, and not for representative exposure assessment in industrial areas.

Some important elements in a measuring strategy for general exposure assessment are:
- code systems used for sampling of measurement objects,
- sampling methods,
- analytical methods,
- assessment of exposure time.

Industrial codes, sampling and analytical methods are relatively easy to compare, e.g. between countries, and a comparison or translation can be carried out immediately. This is not the case for work process codes and assessment of exposure time, where comparisons cannot be carried out directly. These matters will now be discussed in detail.

Industrial codes:

The 4-digit ISIC-code[8] is the basis for almost every national code system on economical or industrial activities. The national versions are often expanded to a 5- or 6-digit system, but a reduction back to the common 4-digit basis is always possible. In 1993 a new code system, NACE, was introduced and made obligatory for statistical purposes within the European Community[9]. The basis of the NACE system is a 4-digit system used by all member states. The common 4-digit basis can be expanded in any country to a 5- or 6-digit system. In this way, comparison between industrial groups is made possible.

Sampling methods

Occupational measurements of gaseous pollution are carried out by adsorption of the pollutant to some suitable matrix, either with the use of a pump or by diffusion; measurements of particulate matter are done by the filter method. Most occupational measurements are carried out by these methods. Within each category, gaseous or particulate, different devices are used, and much scientific work is done in order to compare the different devices[10,11]. It can therefore be concluded that sampling methods are comparable.

Analytical methods

Comparison between analytical methods and/or laboratories is carried out regularly, and all 5 database administrators participate in inter-calibration programmes. However, attention should be paid to any change in analytical methods. For example, improvement of desorption efficiency in GC-analysis will cause a change in reported concentration; recently it has been discovered that the desorption efficiency, used in GC-analysis, for ethyl-benzene differs from that of xylene; ethyl-benzene is part of technical xylene. Generally, these problems can be solved in inter-calibration programs.

Work process codes (and job codes)

It is the experience of the working group that job codes based on education are not suitable for exposure assessment. On the work process level, several national empirical systems have been developed. Because they have developed separately from the experience of each country, they are difficult to compare between countries, and an international system based on a sound theoretical basis is needed. A general work process code system was presented at the above-mentioned database conference[12], but it's suitability to classify work processes from an occupational hygiene viewpoint has not yet been tested in the field. In future, classification of the work process will become more and more important, because workers will be employed to perform certain work tasks (or processes) instead of certain occupations: the worker is employed as a TIG-welder and not as a plumber or a locksmith. This underlines the need for a international code system for work processes.

At the moment, job level codes reflect the education of the person. This is because these codes are adapted from the ILO-system[13]. It may be assumed that this level is of minor importance compared to the industry and work process levels, because of recent developments in multi-skilling of workers. As an example, the occupation or education of the worker adds no further information to the classification mentioned below:
- locksmith or plumber   :welding at shipyard
- unskilled worker or carpenter   :sawing at sawmill

This is only a crude example, and there might be situations where the job level is relevant.

Assessment of exposure time

Exposure time is defined as that fraction of the working day or working shift during which the employees are exposed to non-zero concentrations of air pollutants. The concentration level during the exposure period is set equal to the measured concentration during the sampling period. In general, the exposure time cannot be calculated directly.

The exposure time is used by the occupational hygienist or database holder to estimate full shift exposure. If sampling is performed for a period of 2 hours during the exposure period, the result being 5.0 mg/m^3, and the exposure period is 6 hours, then full shift (8 hours) exposure is calculated as:

$$\frac{6\ hours}{8\ hours} * 5.0 mg/m^3 = 3.75 mg/m^3$$

Sampling time can be a fraction of the exposure time (sampling period embedded in exposure period) and vice versa; sampling time may be part of exposure time (sampling period is overlapping exposure period); or sampling period may be outside exposure period. Often, sampling is performed only during part of the working day, and both exposure time and period are estimated by the occupational hygienist performing the sampling. The estimate is made as a professional judgment.

The difficulties in estimating full shift exposure, without exact knowledge of exposure time, may be solved in two ways. One possibility is sampling during the full shift, and another is sampling at a random time during the shift. The first option is time-consuming, but gives exact information on full shift exposure, relevant for comparison to full shift exposure limits. The second option, a less expensive alternative, gives unbiased estimates of full shift exposure for groups of workers; but this method gives poor estimates of full shift exposure for individual workers.

Obviously exposure time is an important quantity in occupational hygiene, and international consensus on assessment of exposure time is thus required.

PRESENTATION OF XYLENE RESULTS

As an introduction to the comparison of xylene results in this section, here is the method for data reduction. The data reduction was performed in each database before to the interchange of measurements.

In this project the working group defined exposure measurements as measurements with a sampling time longer than 60 minutes and shorter than 8 hours. Data were selected only if the sampling time was within this interval. In each base an integrity check was performed in order to exclude measurements which did not belong to an exposure situation (test of new methods, inappropriate sampling devices or wrong units).

The measurements were performed as personal sampling, personal related sampling or stationary sampling. Both charcoal tubes and diffusive samplers were used as sampling devices, according to the practice in each country. It was not possible to find an intersection of satisfactory size for these different sampling methods. Therefore the comparison will be interpreted as a comparison between exposure assessments, and not as a comparison between measurements.

The comparison between the xylene exposure assessments is carried out as a comparison of geometric means (GM) (relative concentrations in percentage) together with geometric standard deviations (GSD). Concentrations in mg/m^3 will not be reported because some data are private property and confidential.

The GSDs and relative GMs are calculated for all xylene measurements in the bases: measurements within the wood industry, measurements during spray painting and intersection between wood industry and spray painting. The number of measurements and the span of years are also presented in the tables.

In table 1, full shift exposure is presented under the assumption that exposure time equals 8 hours: concentration outside the sampling period is set equal to concentration within sampling period. This means that full shift exposure is set equal to exposure during the sampling period.

In table 2, full shift exposure is presented under the assumption that exposure time equals sampling time, i.e. concentration outside the sampling period is set to zero. Actually, full shift exposure is calculated as measured concentration multiplied by that fraction of the working day which the sampling time consists of.

Notice the meaning of index b in table 1 and index d in table 2;
the GMs under these indexes are calculated as if each of the GM total values were 100%. If, for example, the results from the five wood industries were compared, they would have to be multiplied by the GM total value: the result from the Norwegian wood industry is 104% multiplied by 0.33 equal to 34%. This result can be compared to the Danish result 84%.

Except for the Norwegian result, the GMs in table 1, total, show little variation. As discussed earlier, the variations may be due to differences in:
- industrial structure,
- measurement policy,
- sampling strategy.

The consequences of these differences between countries are impossible to quantify.

On the other hand, important results can be deduced from the wood industry and spray painting measurements:
- The wood industry COLCHIC results are lower than the overall COLCHIC results, while the wood industry results for the rest of the bases are at the same level as the overall results.
- The spray painting results are elevated compared to the overall results (except for COLCHIC and MEGA) and compared to the wood industry results.
- The results on spray painting within the wood industry are lower than the overall spray painting results, but are at the same level as the wood industry results.

Furthermore, the low Norwegian total result in table 1 may be explained either by the analytical routine, where xylene found in white spirit is stored as xylene and not as a fraction of white spirit, or by the prolonged Norwegian sampling time (see page 32).

The ratio between the total columns in table 1 and table 2 reflects the overall fraction of sampling time within the length of the working day for each of the bases. The results in table 2 compared to the results in table 1 show significant differences only in the total column. This is interpreted as meaning that sampling time influence on the results is not dependent on selection of sub-populations. This comparison between table 1 and table 2 is not valid for the NEDB results, because the 2 datasets (in table 1 and 2) originate from different measurements. Notice that the Norwegian average sampling time is twice the average sampling time of the DK, F, and D base.

In general all the reported GSDs are very large. A decrease of the size of the GSDs should be expected for the wood industry and spray painting (and spray painting in wood industry) results compared to the overall results. This is only found for ATABAS and MEGA.

Although the GSDs are large, significant differences (95% level) are found between the GMs of spray painting in wood industry. The French result is significantly lower than the other results. Because the datasets are unbalanced and have unequal variances, intervals of confidence have been used to test differences, instead of analysis of variance.

Table 1. Comparison between exposures to xylene, when exposure time equals 8 hours, by database, for all measurements (total), wood industry, spray painting, spray painting in wood industry (S in W).

Xylene	Total			Wood industry			Spray painting			S in W			Year
	GM %a	GSD	n	GM %b	GSD	n	GM %b	GSD	n	GM %b	GSD	n	
ATABAS	100	5.9	1039	84	5.3	544	192	4.8	300	105	4.2	148	1983-1989
EXPO	33	6.0	2219	104	5.4	244	278	5.7	471	126	6.4	103	1985-1992
COLCHIC	69	6.5	1516	44	5.5	242	105	10	133	37	4.7	46	1987-1993
MEGA	78	7.0	10197	85	4.9	1928	117	5.3	2702	89	3.9	924	1981-1993
NEDB*, job	105	7.4	771	139	7.8	57	185	7.8	164	-			1985-1992
process	-			-			220	7.4	106	-			

total: all xylene results in the databases
GM: geometric mean of concentrations
GSD: geometric standard deviation
n: number of measurements
%a: percentage of ATABAS GM (total)
%b: percentage of own GM (total)
*:   sampling was performed during the whole working day for these NEDB-data

Table 2. Comparison between exposures to xylene, when exposure time equals sampling time, by database, for all measurements (total), wood industry, spray painting, spray painting in wood industry (S in W) in combination.

Xylene	Total			Wood industry			Spray painting			S in W			Year
	GM %c	GSD	n	GM %d	GSD	n	GM %d	GSD	n	GM %d	GSD	n	
ATABAS	34	5.6	1039	89	4.8	544	186	4.9	300	104	4.2	148	1983-1989
EXPO	59	6.0	2219	113	5.3	244	275	5.4	471	138	6.3	103	1985-1992
COLCHIC	23	6.9	1516	46	6.1	242	108	11	133	38	5.4	46	1987-1992
MEGA	23	7.0	10197	100	7.0	1928	120	5.4	2702	93	3.9	924	1981-1993
NEDB**, job process	190 -	7.8	1984	100 -	7.4	132	124 137	7.8 7.8	359 361	- -			1985-1992

total: all xylene results in the databases
GM: geometric mean of concentrations
GSD: geometric standard deviation
n: number of measurements
%c: percentage of own GM (total), when exposure time equals 8 hours (table 1)
%d: percentage of own GM (total from this table)
**:   sampling was performed only during actual exposure for these NEDB-data

REPRESENTATIVENESS OF XYLENE RESULTS

One way to deal further with assessment of exposure time is to carry out a comparison between datasets originating from distinct sampling strategies, particularly when one dataset is from a representative study and the other is from a general control strategy. In this case it is possible to quantify the bias introduced in the control measurement dataset. The bias in the very large datasets of control measurements is discussed intensively here, because there is an objective need for knowledge on workers' exposure to air pollution when standard settings are discussed.

In 1988 the Danish Working Environment Service conducted a representative cross-sectional survey on air pollution in the Danish wood and furniture industries[14]. The survey was designed so that sampling was unbiased according to enterprises, workers and exposure time. This dataset may therefore serve as a representative dataset.

When the measurements from the representative study are isolated from the Danish wood working results, a comparison may be carried out:

	GM %	GSD	n
control meas.: exp. time = samp. time	86	4.5	417
control meas.: exp. time = 8 hours	309	4.5	417
representative study	100	3.8	127

The GMs are normalized according to the representative study. The comparison demonstrates that when the assumption: exposure time equals sampling time is used, exposure is only slightly underestimated (a factor of 0.86).

This exercise quantifies the bias between the Danish overall strategy for control measurements and a representative exposure assessment. The reported bias includes contributions from biased sampling periods, selection of persons and selection of factories. It is a reasonable assumption that the control strategy within the wood and furniture industry does not differ from the control strategy on xylene used in other industries. Thus the present result, on the size of the bias between a representative study and a specific exposure assessment (based on control measurements), may serve as part of a basis for exposure assessments in standard settings or epidemiological studies.

DEFINITION OF CORE INFORMATION

In this project, core information in the bases is defined as that information which is capable of classifying the measurements or distribution of measurements in such a way that they can be used for exposure assessment for certain categories of workers, and in such a way that this assessment may be compared between countries.

In all the bases, different types of information are connected to each sampling or analytical result, such as identification of the person who was measured, sampling method (personal or stationary sampling) and other demographic or administrative information. All these pieces of information are important when the bases are used by the authorities. But in order to prevent all information in the bases being defined as core information (which would make the concept of core information meaningless), we must establish a narrower definition of the term 'core information'.

As an example, the name or any identification number of a person measured is most important when measurements within one person have to be pooled; but this information may be looked upon as a tool, and not as core information. Furthermore, each analytical result, with its sampling time, is essential information; but it would be unnecessary to define this as core information.

From the discussions in the working group, 2 categories of information are identified and defined as core information:

- code systems,
- exposure time.

Together with sampling strategy, these 2 categories of information should be analyzed thoroughly in future joint activities on exposure databases.

RECOMMENDATIONS

During the present project, concrete co-operation among European exposure database administrators was initiated. The first preliminary comparisons have been carried out and scientific results have been reported.

It is recommended that the present exercise is repeated on other substances in order to carry out more detailed comparisons and discussions on code systems and exposure time assessment.

More specifically-used substances should be selected in a new project. Xylene was selected for the present project because it is generally used. Selection of specifically-used substances would bypass some problems with code systems, because coding on industry level would be synonymous with substances: Examples are:

- styrene in glass-fibre reinforced plastic production,

- lead in battery production,

- perchloroethylene in cleaning of clothes.

REFERENCES

1. Smith, M.H.P.; Glass, D.C.: *The Availability of Occupational Exposure Data in the European Communities.* EUR 14378 EN. Commission of the European Communities, Luxembourg (1992).

2. Vinzents, P.S.; Carton, B.; Fjeldstad, P.; Rajan, B.; Stamm, R: *A preliminary comparison of exposure measurements stored in European databases on occupational air pollutants and definition of core information.* Submitted to Appl. Occup. Environ. Hyg. (1993).

3. Carton, B.; Fjeldstad, P.; Rajan, B.; Stamm, R; Vinzents, P.S.: *A comparison of European exposure databases for compatibility and common standards.* British Occupational Hygiene Society Conference (1994).

4. Rappaport, S.M.: *Assessment of long-term exposures to toxic substances in air.* Ann. Occup. Hyg. 35(1): 61-121 (1991).

5. Heederik, D.; Hurley, F.: *Occupational exposure assessment: Investigating why exposure measurements vary.* Appl. Occup. Environ. Hyg. 9(1): 71-73 (1993).

6. Rappaport, S.M.; Kromhout, H.; Symanski, E.: *Variation of exposure between workers in homogeneous exposure groups.* Am. Ind. Hyg. Assoc. J. 54(11): 654-662 (1993).

7. European Standardization Commitee (CEN): *Workplace atmospheres - Guidance for the assessment of exposure to chemical agents for comparison with limit values and measurement strategy.* prEN 689. Brussels (1992).

8. United Nations: *International Standard Industrial Classification of all Economic Activities.* Statistical Papers Series M, No. 4, Rev. 3. UN, New York (1990).

9. Official Journal of the European Communities: Council Regulation (EEC) No 3037/90 of 9 October 1990 on the *Statistical Classification of Economic Activities in the European Community.* 33 L 293. Commission of the European Communities, Luxembourg (1990).

10. Pagnotto, L.D: Gas and vapor sample collectors. In: *Air sampling instruments.* 6th. ed. ACGIH. Ohio (1983).

11. *Personal communication*, Lee Kenny, HSE.

12. Burr, H.: *Presentation of a work process code that enables data analysis pooling different types of knowledge.* Submitted to Appl. Occup. Environ. Hyg. (1993).

13. International Labour Office: ISCO-88 *International Standard Classification of Occupations.* ILO, Geneva (1990).

14. Vinzents, P.S.; Laursen, B.: *A National Cross-sectional study of the working Environments in the Danish Wood and Furniture Industry - Air Pollution and Noise.* Ann. Occup. Hyg. 37(1): 25-34 (1993).

European Foundation for the Improvement of Living and Working Conditions

Exposure Registers in Europe

Luxembourg: Office for Official Publications of the European Communities, 1994

1994 – 42 pp. – 16 x 23.5 cm

ISBN 92-826-8737-6

Price (excluding VAT) in Luxembourg: ECU 7